EARTH ADVENTURE

WRITTEN AND ILLUSTRATED BY
STU DUVAL
COVER DESIGN BY
PAUL XU
ART DIRECTOR:
PAUL XU
EDITOR:
JON THIBAULT
MOTHER NATURE LTD.

Copyright © 2008 by Mother Nature Ltd.

1411 4ᵗʰ Avenue, Suite 765
Seattle, WA98101

All rights reserved. No part of this
publication may be reproduced, stored in a
retrieval system, or transmitted by any
means, electronic, mechanical,
photocopying, recording or
otherwise, without the prior permission
of the publisher

ISBN-13：978-0-9814547-3-3
ISBN-10：0-9814547-3-9

Printed in Taiwan

CHARACTERS

Lily
Twelve year-old with a project to complete.

Buz
Protector of the earth and Lily's tour guide.

Basil
A hapless forest gnome.

Mother Earth
Buz's Mom-- the personification of nature.

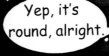
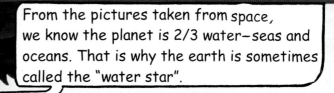

The models of the earth are round, because the radius of the equator differs only slightly from the radius of the rotational axis.

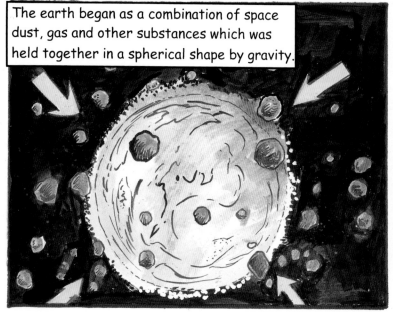

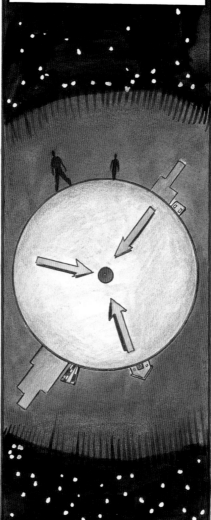

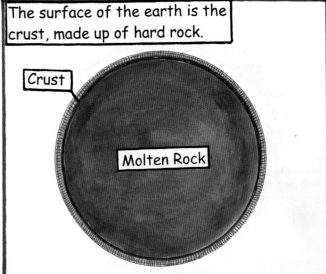

From the earthquake research, we can estimate that the temperature at the earth's core is over 5000°C.

5000°C

Rock melts at 1000°C.

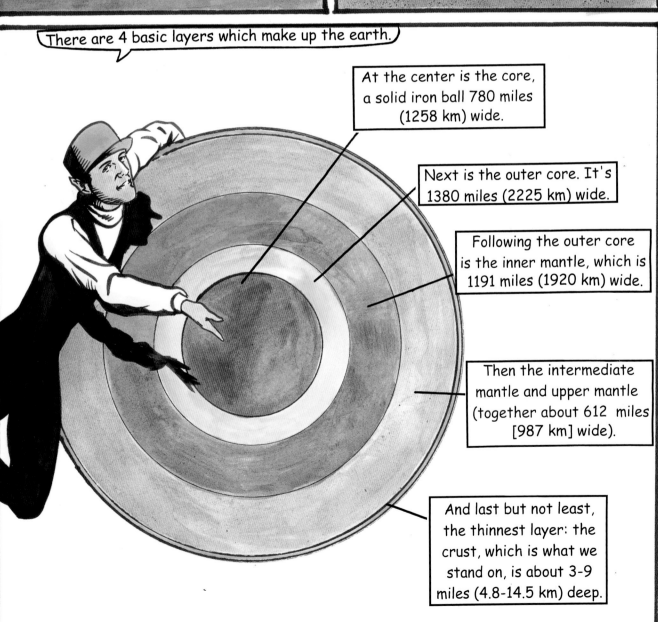

There are 4 basic layers which make up the earth.

At the center is the core, a solid iron ball 780 miles (1258 km) wide.

Next is the outer core. It's 1380 miles (2225 km) wide.

Following the outer core is the inner mantle, which is 1191 miles (1920 km) wide.

Then the intermediate mantle and upper mantle (together about 612 miles [987 km] wide).

And last but not least, the thinnest layer: the crust, which is what we stand on, is about 3-9 miles (4.8-14.5 km) deep.

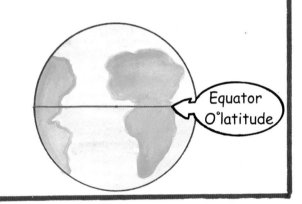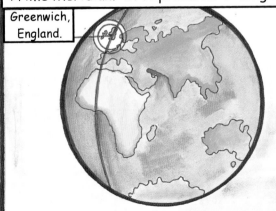

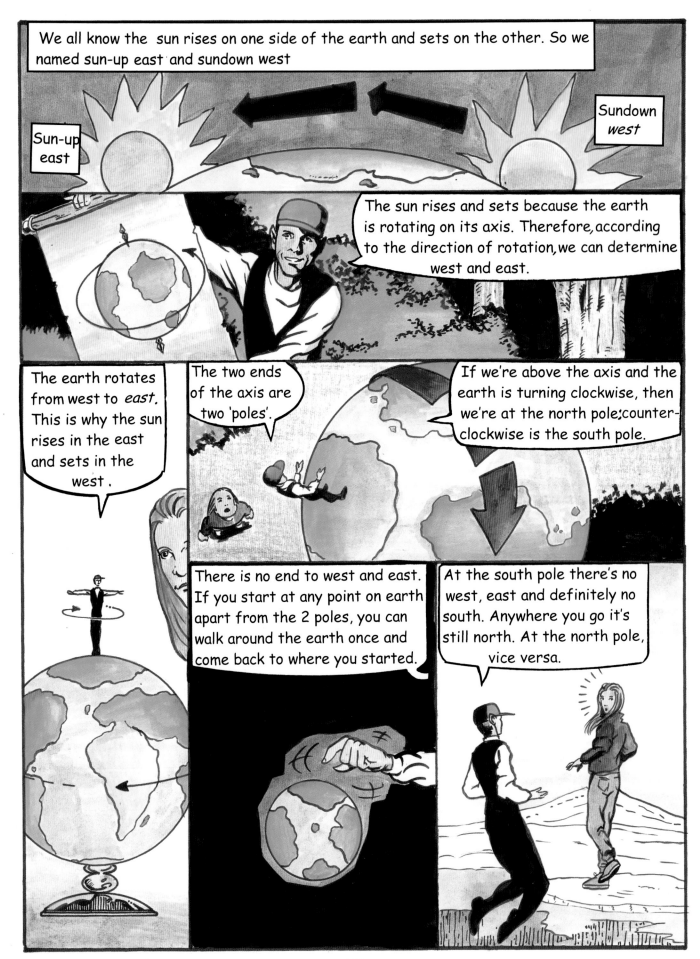

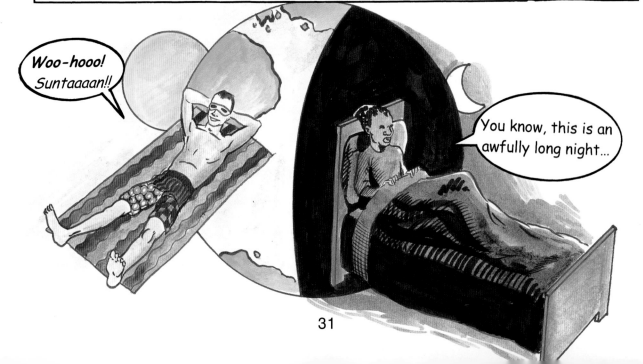

The sun's gravity keeps the earth in orbit.

Stay right where you are.

I'm not moving, pal.

The planets don't easily go out of orbit because the sun keeps 'em in place with mass gravity.

The sun's gravity keeps the earth in orbit. According to Newton's law, a moving object would travel in a straight course if there were no outside interaction. In other words, the earth would continue moving in a straight line if there was no sun and no mass gravity.

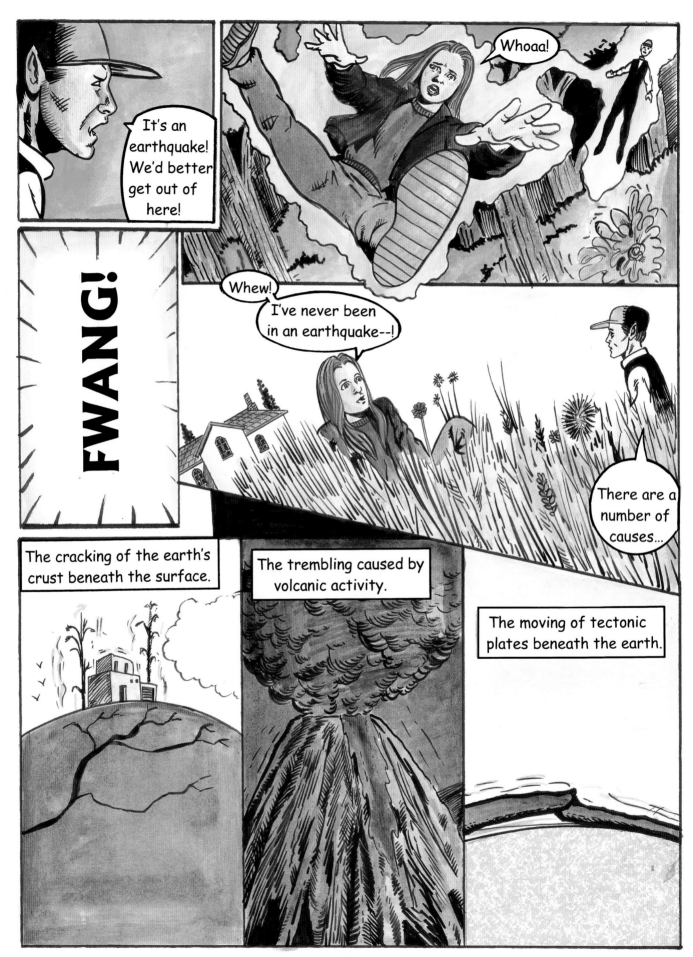

Some areas often have earthquakes because of the cracked crust. Some never experience earthquakes at all, because there is no crack in the crust.

Hey, we're at another volcano--!

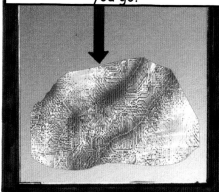

A volcano begins to form when rock melts about 60 miles (96 km) beneath the crust. The temperature rises the deeper you go.

There are many mountains on the crust. Lava is often stored beneath these mounds, as there is less pressure.

Lava seeps through the cracks in the mountain and when the pressure reaches an unbearable level, the magma erupts out of the mountain.

The hole left at the top of the mountain is called the 'crater'. The magma gathers and eventually cools around the crater.

The volcano we see now is the result of the magma condensed around the crater in the cool atmosphere.

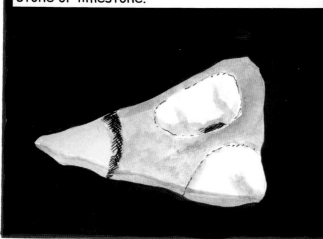

Travellers in the 15th century wanted to explore more of the world, but needed something easier to use than wooden rowboats.

Sails were invented, and with their use, studies of the wind have produced 12 levels of wind strength.

Wind level 0.
Velocity 0.4 m/sec.
Smoke rises straight up, water surfaces are smooth, and leaves don't move.

Wind level 1
Velocity 0.4-1.4 m/sec.
Smoke wisps with the wind, but the anemometer does not move.

Wind level 2
Velocity 1.6-3.3 m/sec.
The wind can be felt, the leaves wave slightly, and the anemometer turns.

Wind level 3
Velocity 3.4-5.4 m/sec.
The anemometer turns, leaves rustle, flags flap and small ripples appear on the water.

Wind level 4
Velocity 5.5-7.9 m/sec.
Tree branches shake, dust, paper & sand starts flying.

Wind level 5
Velocity 8.0-10.7 m/sec.
Small trees bend. At sea, sails become unnecessary and small waves appear on the water.

Wind level 6 Velocity 10.8-13.8 m/sec.
Power lines clash, it becomes hard to hold an umbrella and big trees sway.

Wind level 7 Velocity 13.9-17.7 m/sec.
It becomes hard to walk against the wind, there are big waves in the sea and everything is shaking.

Wind level 8
Velocity 17.8 -20.7 m/sec.
We feel strong resistance, branches break off and huge waves are created.

Wind level 9
Velocity 20.8-24.4 m/sec.
Chimneys collapse and damage is caused to small houses.

Wind level 10
Velocity 24.5-28.4 m/sec.
Trees are uprooted and roof tiles shift.

Wind level 11 Velocity 28.5-33.5 m/sec.
Ferry cruisers capsize and very violent waves begin churning the sea.

Wind level 12 Velocity above 33.5 m/sec.
This very rarely occurs on land. Huge tidal waves can be fatal for ships.

"I feel a breeze."

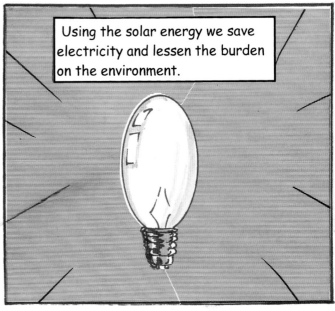